UNBELIEVABLE FACTS ABOUT BEARS FOR KIDS

Amazing, Interesting, and Fun Trivia You Need to Know About This Cool Animal with Quiz Questions.

Kids Intelligentsia

Alpha Zuriel Publishing

Copyright © 2021 by Kids Intelligentsia

The content contained within this book may not be reproduced, duplicated, or transmitted without direct written permission from the author or the publisher.

Under no circumstances will any blame or legal responsibility be held against the publisher, or author, for any damages, reparation, or monetary loss due to the information contained within this book; either directly or indirectly.

Legal Notice:

This book is copyright protected. This book is only for personal use. You cannot amend, distribute, sell, use, quote, or paraphrase any part, or the content within this book, without the consent of the author or publisher.

TABLE OF CONTENTS

INTRODUCTION 4

PREHISTORY 6

CHARACTERISTICS OF BEARS 8

TYPES OF BEARS 10

THE BEAR AND EVOLUTION 21

WHAT DO BEARS EAT? 28

FACTS ABOUT THE POLAR BEAR: 33

CONCLUSION 36

QUIZ 37

BONUS PAGE 40

INTRODUCTION

BEAR PROFILE

Kingdom: Animalia

Phylum: Chordata

Class: Mammalia

Order: Carnivora

Family: Ursidae

Genus: Ursus

Even without being great specialists in bears, we all know that bears remain very large, robust, powerful, intelligent, greedy quadrupeds, able to stand on their two hind legs to impress their rivals or observe danger.

We also know that these animals are rather solitary, opportunistic and that they adapt their food, their existence according to the weather, the seasons, and the countries. Some well-known bears, such as the grizzly, will overwinter rather than hibernate, while others, such as the sun bear, will remain active year-round.

Can we, however, say that we know these creatures? If they are fascinating, intriguing, powerful, dangerous, and inoffensive. To better understand these animals, it is essential to study them, discover them, and study their history.

PREHISTORY

In Europe, humans and bears have lived together for over 400,000 years. Thus, Neanderthals and Homo sapiens rubbed shoulders with brown bears and cave bears on a daily basis.

Unfortunately, it is difficult to determine the nature of their relations due to the rarity of archaeological traces. Archaeologists have found among the dwellings or caves many tools, bones, teeth, and ornaments from the bodies of ursids.

Is the bear a god or a direct competitor of men in the struggle for survival? The question remains.

CHARACTERISTICS OF BEARS

Bears are very large and robust animals. Depending on the species, its weight and size vary. The largest bear species is known as the polar bear, which can measure up to almost 3 meters and weigh up to 700kg. In most species, male bears are larger than females.

Their eyes, ears, and tail are small. Its limbs are short. They have a large fur, which varies in color according to the species (they can be black, brown, white, and with spots). They have huge claws with which they capture their prey.

Bears are animals with magnificent eyesight, comparable in nuances to humans. Their smell is also great, and their hearing is similar to that of dogs.

They are calm animals that move slowly, but if they need to run, they can do it at high speed. They are most active during the afternoon - night, but are also active during the day.

Also, they are pretty lonely. Only females with their young cubs are usually seen in groups.

TYPES OF BEARS

THE BEARS PROPER (URSUS)

The genus Ursus has been subdivided into two sub-genera on secondary characters: Ursus (Brown bears), which has many subspecies, and Euarctos (Black bears). The cave bear (Ursus spelaeus), now extinct, was related to the present brown bear. Two other extinct species, Ursus deningeri and Ursus etruscus, also belonged to this group.

THE BROWN BEAR

The brown or European bear (Ursus arctos) inhabits the mountainous and wooded regions of Europe and northern Asia, from Norway to Spain and from Northern Siberia to the Himalayas. To Afghanistan, China, Tibet and is found in northern Japan. It is brown in color, quite often varying from white to the neck and throat, but the colors of its coat and its dimensions vary greatly depending on the regions it inhabits.

THE ALPINE BEAR

The Alpine brown bear wears a white collar at a young age which fades of as it ages. It is only found in the mountains' wildest regions, where the caves and the trunks of old oaks and hollow beeches serve as a retreat. It feeds on buds, fruits, fungi, roots, leaves and occasionally ravages wheat fields and vines in summer. He searches for bee nests, and to find them, it climbs trees and eats honey. It searches ant hills for eggs and larvae. Older individuals are more carnivorous and hunt small animals, chase games, prowl around pastures to catch a sheep or young calf. In winter, they are not afraid to enter the stables through a breach in the roof: if they can slaughter a cow, their strength is great enough to carry the corpse by the same path and drag it from a distance to eat.

AMERICAN BLACK BEAR

It is the most abundant bear in North America and considered one of the most intelligent mammals in the world. Not for nothing was he captured a lot for circuses. 16 subspecies of American black bears have been said to live in forests and mountains, although the color of their fur varies between them from black to white, with different shades.

Males can be up to 2.80 meters long; females, up to 2.55 meters. Although they are quite large and heavy, they have great ability to climb and run; bears can run a speed of up to 55 kilometers per hour. Although you'll most likely see them alone, it is possible to find them gathered in areas where there is a lot of food.

GRIZZLY OR FEROCIOUS BEAR

(Ursus horribilis) is a large species, a brownish-gray and remarkable for its long nails. The western territories of the United States (Montana and Wyoming) and Canada and the Anglo-Americans portray him as very fearful when he is in the presence of humans. Baird's Ursus horriaeus which inhabits the southern slopes of the Rocky Mountains, Sonora, California and Mexico, and Ursus alascensis (Merriam) from southern Alaska are only subspecies.

ASIAN BLACK BEAR

It lives in the forests of Asia, from Iran to Japan, and can even be found in Taiwan. Its closest relative is the American black bear and shares a common ancestor with the brown bear and the polar bear. It is omnivorous and loves berries, fruits, nuts, as well as honey, fish, and some small animals. It also consumes carrion.

Its ecosystems have been invaded in its wide area of distribution, which explains its aggressiveness and possible attacks on humans. Seven subspecies of the Asian black bear have been recognized so far, but it is already extinct in many places, either through habitat loss or hunting.

POLAR BEAR (THALARCTOS)

The Polar Bear (Thalarctos maritimus (Gray), sometimes called Ursus maritimus) is distinguished by its elongated head and neck, small and narrow molars, the soles of the feet are hairier than in other species. Its color is a uniform dirty white, and the lining of the mouth is a purplish-blue. It inhabits the arctic regions of both continents, living almost constantly in the midst of ice, swims easily, and feeds on Seals, Reindeer, Blue Foxes, and Fishes. It is a formidable adversary, especially in winter when the sea is taken, and it is more difficult to find food; starved by several days of fasting, it attacks humans themselves. The full female spends the winter alone in a hole dug in the snow, the narrow chimney of which, formed by the heat of her breathing, often betrays the presence; this is where she usually gives birth to two cubs.

It is the super predator of the Arctic and an excellent swimmer. In recent years, it has become the symbol of climate change. Their home, the northernmost area of the planet covered by ice, is at constant risk from rising temperatures.

It has a wider profile than that of other bears, and its legs have been adapted to support itself on ice sheets, from where it propels itself to hunt seals. As with the spectacled bear, in reproduction, the females wait to implant the fertilized ovum, preparing in the previous months by storing as much fat as possible, although the species does not hibernate.

MALAYAN BEAR

It is the smallest of the bears and is found in the tropical forests of Southeast Asia. They easily climb trees to find food, especially coconuts, their favorites.

Although it is omnivorous, it consumes everything from seeds to insects to small mammals. Like the spectacled bear, it does not hibernate, and females can have two cubs a year. These bears are fewer and fewer due to the loss of their habitat caused by agriculture, mining, and timber extraction, as well as illegal hunting.

THE BIG-LIPPED BEAR (MELURSUS)

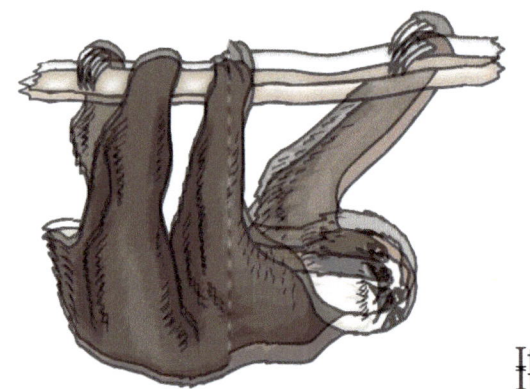

It is the most famous species thanks to the character of Baloo, Mowgli's teacher in 'The Jungle Book', by Rudyard Kipling.

The sloth bear lives in grasslands and forests in India, Nepal, Sri Lanka, and Bangladesh. Its hair is longer and straighter than the rest of his Ursid cousins.

Their feet are curved inward, and they have a prominent snout. Its diet is extensive: eggs, honey, flowers, and tubers, but its weakness is ants and termites.

It reproduces once a year, and its pregnancy lasts eleven months. The population is decreasing due to deforestation and illegal hunting for their skin and bile (used in Chinese medicine).

THE SPECTACLED BEAR

The Ornate Bear or Spectacled Bear (Tremarctos ornatus) of South America is a type of a distinct genus closely related to the Coconut Bear (Helractos). It is small in size and inhabits the Andes of Peru, Colombia, Bolivia, and Chile. The gardener of the forests is the only bear in South America and walks through the Andean mountains, from Venezuela to Bolivia. Their characteristic spots around the eyes, muzzle, and chest are different in each individual. Like the panda and the Malayan bear, it does not hibernate.

They are excellent climbers and their claws are used to grasp tree branches, plant stems, or dig in the ground. Although they are omnivores, they consume mainly fruits and plants.

THE BEAR AND EVOLUTION

In the classification of species, bears belong to the order of carnivores. Likewise, in many cultures, this animal is considered to be cannibalistic and bloodthirsty. However, the vast majority of ursids eat almost nothing but fruits, insects, leaves, roots, fish, honey. You will understand, most bears are omnivorous.

The big panda, however, is only herbivorous. He only consumes bamboo. On the other hand, the polar bear only feeds on meat, mainly seals or various marine mammals. Attacks on human beings are therefore extremely rare. The bear almost always tends to run away from them.

THE ORIGIN OF MODERN BEARS

While many mysteries still need to be unraveled by zoologists, the discoveries of fossilized bears and advances in genetics now provide us with new clues about their history.

Ursidae belongs to an ancient family. Truth be told, the earliest traces of bear-like animals date back almost 35 million years. At this point, their jawbones stand out from other carnivores.

These animals previously lived in Asia but gradually spread to Europe and North America. Indeed, at this time, America remains connected to Eurasia. Bears do not look like the ones we know today since they are more like dogs, coyotes, both in their anatomy and in the shape of their limbs.

Their habitat joins with open spaces, such as meadows or savannahs. Note that today, we do not find bears in Africa. All eight species of bears today belong to the same family. Indeed, their origin is common. Their most distant relatives are called Filholictis or Ballusia.

The Ursidae line then split into three main branches. Among them, the group of large pandas, comprising a small group of closely related species, the short-faced bears, and finally the ursids, which have widely mixed in the northern region, now comprising six species, including the brown bear and the black bear.

TWO EXTINCT BEARS

First, the short-faced bear, also called Arctodus. It was one of the largest bears to have existed: it thus measured 1.80 m in width and 3.50 m standing. It was therefore the largest predatory mammal of the Ice Age! It disappeared about 10,000 years ago at the end of the Pleistocene.

Then, the best known of all: the cave bear, Ursus spelaeus. Some of them reach 1.30 meters in width and could exceed 3.50m in height in an erect position, thanks to long limbs. They were also three times heavier than current brown bears.

MAN AND BEAR

Bears continue to feed the imagination of men through multiple myths, legends, beliefs, festivals. Their human-like nature, their strength, their wild and domesticable side intrigue men all over the world. The bear is at the same time admired, feared, hated, hunted, ridiculed, mocked, and sometimes even rehabilitated.

THE BEAR IN THE MIDDLE AGES

The bear is synonymous with power, strength, authority. It arouses fear, respect, and admiration at the same time. His body is associated with that of warriors, powerful men. Its image is also used to establish honor and power. At the beginning of the

Middle Ages, the bear was treated with great honor. Its presence in the legend of King Arthur proves it. Arthur is strong as a bear. His name Arthur comes from "arth" which means "bear" in Celtic. He is the perfect ruler, strong, intelligent, and courageous.

THE BEAR IN MODERN TIMES

With the advent of modernity, the bear became a ridiculed animal.

During the French Revolution in 1793, street performances were banned in the capital. All animals potentially dangerous to humans, such as bears, are kept at the Jardin des Plantes. This legislation led to the creation of the future wildlife display.

In the 19th century, a bear farming school was established in Ariege. In the 20th century, bears gradually lost their trait of being wild and dangerous animals. The bear population starts declining.

The creation of the Teddy bear plush marked a turning point in modern times. The bear's image becomes positive again: it becomes an affectionate animal par excellence, friendly, to which kids become attached.

THE BEAR AND PARTY

We find the bear in many ceremonies or beliefs. The Haidas, people of western Canada, revere the animal. The bear is the "brother of man". It must thus be represented on totem poles. "The spirit of the bear" is therefore an integral part of the culture of the Plains Indians.

The Ainu, in northern Japan, follow the "bear ritual". This celebration consists of ritually sacrificing an adult bear, raised by men so that its protective spirit comes to guarantee the success of men in the hunt. Through these rituals, Man must both attract the bear's attention and appease his anger.

Both sacred and coarse, the bear symbolizes the beginning of winter as much as the return of spring. Therefore, it can be used as a message of fertility and as an omen of evil, lust, and gluttony. Even today, carnivals are organized, as in Romania, where the bear is ridiculed.

WHAT DO BEARS EAT?

Despite what is commonly believed, bears are not carnivores. Most of the species are omnivorous; they eat fruits, insects, small mammals, fish, honey, nuts, seeds, grasses, eggs, etc. That is why their teeth over the years have evolved to adapt to their varied diet.

The panda bear is almost 100% vegetarian (herbivore) since it basically eats bamboo. However, it can also eat some insects, fruits, fish, and small mammals.

The polar bear is carnivorous since it lives in a very cold climate; its only source of food is the fish and mammals that inhabit the Arctic.

Most bears eat much more in the summer to store fat for the winter. They do this to have more reserves during the hibernation season.

Like many other animals, they seek the humid areas where there is water nearby to establish themselves. This way they know that they will not lack food since water is a sign that there is life (both plant and animal).

HIBERNATION OF BEARS

Most bears hibernate in winter. This means that they retreat to a cavern, cave, or refuge where they sleep while the cold lasts. Bears can hibernate from a few weeks to several months.

They can decide, in the event that they find food and the climate is not very harsh, not to go into hibernation. Hibernating is the bear's way of surviving when food is scarce.

During hibernation, bears have a slower heartbeat and their body temperature slightly lowered. They do not eat, drink, or relieve themselves.

In some species only pregnant females hibernate. They give birth during this period of hibernation.

The species of bears that do not hibernate are: The panda bear due to its herbivorous diet and the sloth.

THE REPRODUCTION OF BEARS

The breeding season for bears runs from May to July. Females belonging to smaller bear species reproduce earlier than larger ones. Two cubs are usually born at each birth, but bears can born up to six cubs at each birth.

BABY BEARS ARE CALLED CUBS

At birth (usually between the months of January and February), the cubs have neither hair nor teeth. And they stay with their mother for about 3 years. During this time, the mother bears are very dangerous, as they become more aggressive to protect their little ones. They are very good mothers.

BEAR THREATS

Most Bears are found at the top of the food pyramid. This means that it is not threatened by other animals. Although baby bears are vulnerable, they can be attacked by big cats and wolves.

Unfortunately, most bear species are in danger of extinction today. Because of the humans who hunt them to obtain their skin, they destroy and occupy their natural habitat. Due to the human way of life, the bears are affected by climate change.

FACTS ABOUT THE POLAR BEAR:

- Canada is home to nearly 60% of the planet's polar bears.
- Polar bears can swim up to 100 kilometers! Watch out for prey!
- The Canadian two-dollar coin features a polar bear.
- The fur and skin are oily and very water repellent, so they can easily drain away water and dry very quickly.
- Polar bears have such a powerful thermal insulation system that they sometimes overheat; they must then swim in the icy waters to cool down.
- Polar bears use the floating ice sheets (pack ice) as a hunting platform to catch seals and other marine mammals on which they feed. Without these patches of ice, it is difficult for them to eat properly.
- The female polar bear is twice the size of the male, weighing between 200 and 300 kg. The male polar bear reaches its adult size around 8 to 10 years old while the female stops growing around 5 to 6 years old.
- A polar bear can eat up to 46 kg (100 lbs) of food in one go.
- A Polar bear can live up to 35 years in the wild.

- The polar bear is an animal that can interbreed with the grizzly. The crossing of the 2 species gives birth to a hybrid called "pizzly or grolar. Unlike other hybrids, the "pizzly" can breed.

- In summer, polar bears must return to dry land. Unable to hunt seals, they live mainly on their fat reserves. They must conserve their energy by remaining inactive more than 80% of the time.

- A polar bear weighing 1600 kg and standing at least 3 meters has been discovered! He lived 500,000 to 2 million years ago. This giant, short-faced bear would have been the largest and most powerful meat consumer of its time, scientists say.

CONCLUSION

While the population of some bears, such as the black bear comprising 900,000 individuals around the world, are doing well, the majority of the bear groups are endangered. Man is continually inspired by animals and never stops wanting to understand their history.

QUIZ

1. Brown bears have a scientific name, what is it?
   _____
   _____

2. Are bears carnivorous or omnivorous?
   _____
   _____

3. Bears attack threats by throwing rocks. 'True' or 'False'?
   _____
   _____

4. The grizzly is a subspecies of the Asian black bear, 'True' or 'False'?

5: The 'bear ritual' is performed in what country?

6: Bears teeth have become sharper over the generation. 'True' or 'False'

7: Name the type of bear that gestates for 11 months.

8: Name one of the two extinct bears discussed in this book.

9: For how long have bears been living with humans?

10. In the middle ages, what does the bear symbolize?

BONUS PAGE

COLOR

www.ingramcontent.com/pod-product-compliance
Lightning Source LLC
Chambersburg PA
CBHW051932210526
45473CB00006B/2221